The MOON

Robin Kerrod

Lerner Publications Company • Minneapolis

This edition published in 2000

Lerner Publications Company
A Division of Lerner Publishing Group
241 First Avenue North, Minneapolis MN 55401 U.S.A.

Website address: www.lernerbooks.com

© 2000 by Graham Beehag Books

Library of Congress Cataloging-in-Publication Data

Kerrod, Robin.
    The Moon / Robin Kerrod.
        p. cm.—(Planet library)
    Includes index.
    Summary: Introduces the moon, its properties, formation,
and exploration.
    ISBN 0-8225-3900-4 (lib. bdg.)
    1. Moon—Juvenile literature. [1. Moon.] I. Title.
    QB582.K47 2000   99-34703
    523.3—dc21

Printed in Singapore by Tat Wei Printing Packaging Pte Ltd
Bound in the United States of America
1 2 3 4 5 6 – OS – 05 04 03 02 01 00

CONTENTS

Nearly every night, the Moon rises into the sky and helps lighten our darkness. It outshines the stars, the planets, and all the other heavenly bodies in the night sky. At its brightest, it casts shadows. The Moon also looks much bigger than the other heavenly bodies.

The Moon is actually much smaller than Earth—about as big across as the United States is wide. It looks bigger and brighter than the planets and stars because it is so much closer to Earth. It lies only about 239,000 miles (384,000 km) away. The next nearest body to Earth is the planet Venus, which lies more than 26 million miles (41 million km) away.

The Moon is always there in the night sky. This is because it moves with Earth as Earth travels through space. It circles around Earth in a constant path, or orbit. It takes the Moon about a month to circle Earth, and the word *month* comes from an old form of the word Moon.

The Moon is Earth's only natural satellite. A satellite is a small body that circles around another, larger body. Some planets, such as Jupiter and Saturn, have many satellites, or moons, circling around them.

We know more about the Moon than about any other heavenly body. Astronomers have studied it closely through telescopes for hundreds of years. Space scientists have sent probes to take close-up photographs and land on its surface. And astronauts have traveled to the Moon, walked on its surface, and brought back samples of Moon rocks and soil.

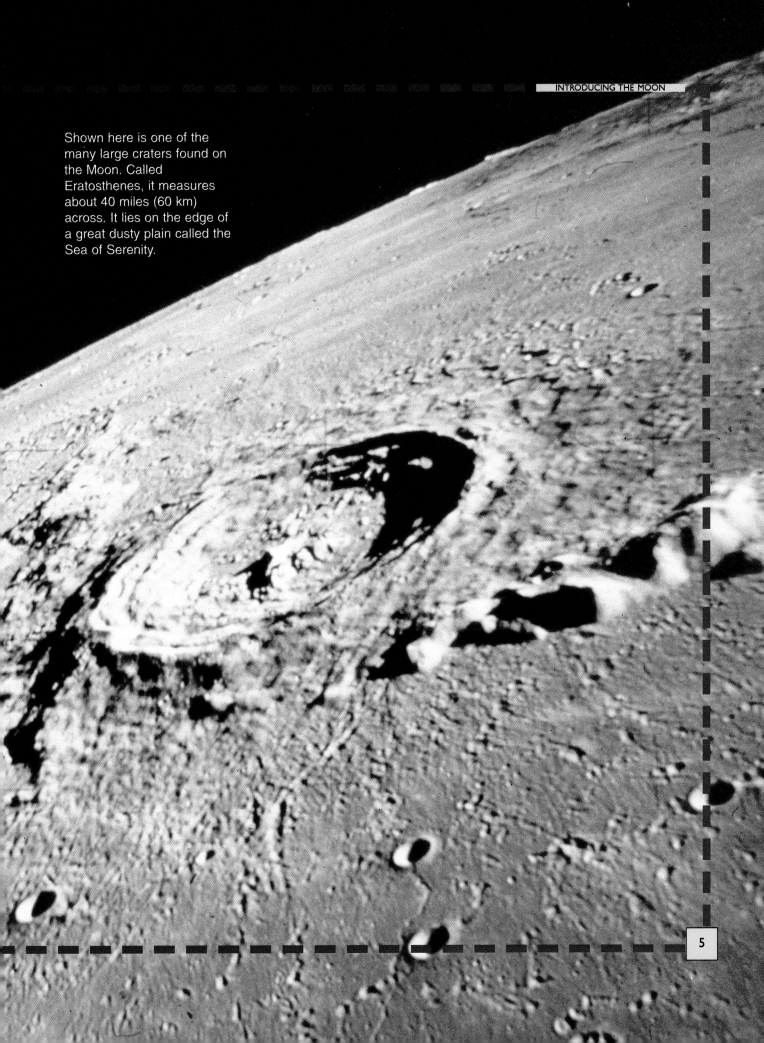

Shown here is one of the many large craters found on the Moon. Called Eratosthenes, it measures about 40 miles (60 km) across. It lies on the edge of a great dusty plain called the Sea of Serenity.

Looking at the Moon

Every month, the Moon circles around Earth. As it does so, it appears to change shape, going from a thin crescent to a full circle and back again.

But the Moon does not really change shape during the month. The changes we see happen because of the way the Moon shines. It does not give out any light of its own. Instead, it reflects, or sends back, light from the Sun.

The shape of the Moon in the night sky depends on how much of its surface we see lit up by sunlight. This changes night by night as the Moon changes its position in space. We call the Moon's changes in shape its phases.

We see different parts of the Moon lit up by light from the Sun at different times. For example, when the Moon is on one side of Earth (position 1), we see its left-hand side lit up. When the Moon is on the other side of Earth (position 2), we see its right-hand side lit.

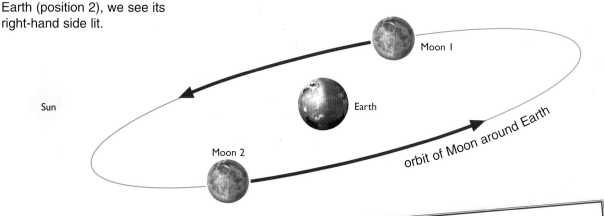

Sun

Moon 1

Earth

Moon 2

orbit of Moon around Earth

Man in the Moon

Long before an astronaut became an actual man on the Moon, people talked about the Man in the Moon. This is because, with a little imagination, you can make out a human face in the features visible on the surface during the full moon. Dark areas form the two eyes and mouth of the Moon. Or maybe you can imagine the face of a Woman in the Moon, or the body of a rabbit with long ears.

PHASES OF THE MOON

During the phase we call the new moon, we cannot see the Moon at all. At this time, the Moon is positioned on the same side of Earth as the Sun. The Sun lights up only the Moon's far side. The side of the Moon that faces us remains dark.

Over the next few days, the Moon starts to become visible as its position changes. At first we see only a region around the edge of the Moon lit up by the Sun. We call this a crescent moon. Gradually, the crescent grows and grows. A week after the new moon, we see half of the Moon's face lit up. It looks like a semicircle. We call this phase the first quarter.

Crescent Moon

Last Quarter

Full Moon

First Quarter

Crescent Moon

A week after the first quarter, we see the whole face of the Moon lit up. This phase is the full moon. The Moon has now traveled halfway around Earth since the new moon. It lies in the opposite direction in the sky from the Sun.

In the days after the full moon, we gradually see less and less of the Moon's face lit up. After a week, we see only half of it. It looks like a semicircle again. This phase is the last quarter. After another week, we can see only a slim crescent. Then the Moon moves back into position between Earth and the Sun and disappears from view. It becomes a new moon once again. The Moon takes 29½ days to go through its phases from one new moon to the next. When it appears to be growing in size from new to full, we say it is waxing. When it appears to be shrinking in size from full to new, we say it is waning.

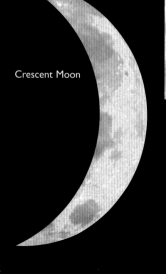

Above: An eclipse of the Moon. The Moon has moved into Earth's shadow in space. But you can still see the Moon faintly. This is because some sunlight still reaches it after passing through the Earth's atmosphere.

EARTHSHINE

During the crescent moon, most of the Moon's face is dark. But it is not completely dark. If you look carefully, you can see that it gives off a pale grayish light. This light has been reflected from Earth and is known as earthshine. A popular name for it is "old moon in the new moon's arms."

IN THE SHADOWS

Just as you cast a shadow on the ground, planets and moons cast shadows in space when they block the Sun's light. Earth casts a shadow, and so does the Moon.

Sometimes the Moon passes into the shadow of Earth. We call this an eclipse of the Moon, or a lunar eclipse. Lunar eclipses happen about once or twice a year. The Sun, Earth, and the Moon line up exactly in space, with Earth positioned between the Sun and the Moon.

A lunar eclipse always takes place during a full moon. As the Moon moves slowly into Earth's shadow, we see less and less of it lit up. After about an hour, the Moon is completely in Earth's shadow. But it does not disappear from view. It still gives off a dim pinkish light.

Moon Madness

In ancient times, people held strange beliefs about the Moon. If you gazed at the Moon for too long, they said, you would go crazy. The word *lunatic*, meaning a crazy person, comes from *luna*, the Latin word for Moon. Another old belief was that at the time of a full moon, some men could turn into savage, wolflike creatures called werewolves, which would spend the night killing people.

The Moon remains in Earth's shadow for more than 1½ hours. Then it takes about another hour to come out of the shadow completely. From start to finish, an eclipse of the Moon can last up to about 3½ hours.

Sometimes the Sun, the Moon, and Earth line up so that the Moon lies between the Sun and Earth. Then the Moon casts a shadow over part of Earth. We call this an eclipse of the Sun, or a solar eclipse. When this happens, day turns suddenly into night. But because the Moon casts only a small shadow that just reaches Earth, a solar eclipse can be seen only from a small part of Earth. It lasts only for a few minutes. Solar eclipses are not as common as lunar eclipses.

NEAR SIDE AND FAR SIDE

The shape of the Moon in the night sky changes all the time. But the features you see on the Moon always stay the same. The face of one full moon looks exactly the same as the face of the next full moon, and the next, and the next. This is because, from Earth, we only see one side of the Moon—the near side. We never see the other side, the far side.

A total (complete) eclipse of the Sun. Here the Moon has passed in front of the Sun and blocked most of its light. But we can now see the outer atmosphere of the Sun. We call this the corona, which means crown.

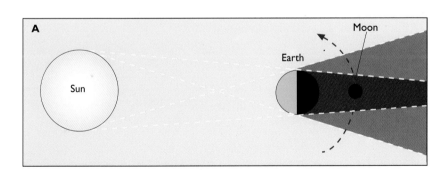

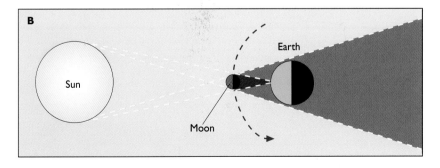

A lunar eclipse (A) occurs when Earth is directly between the Sun and the Moon. A solar eclipse (B) occurs when the Moon is between Earth and the Sun.

The Moon was probably born when small lumps of rock in space began bumping into one another and joining together. The newborn Moon was a ball of red-hot liquid rock. It slowly cooled down and was then bombarded with rocks from outer space. They created many of the Moon's craters. Some of the biggest craters later filled with liquid rock to form great dark plains, called seas.

Birth of the Moon

The Moon is about as old as Earth. We are not sure how or even where it was formed. But we do know how it came to look like it does.

Some of the Moon rocks the *Apollo* astronauts brought back are up to 4.5 billion years old. They are thought to be part of the Moon's original outer layer, or crust. The crust formed not long after the Moon itself. This makes the Moon about the same age as Earth.

Astronomers are not sure about how the Moon was formed, but they have put forward a number of ideas. One is that Earth and the Moon were formed together from lumps of rocky material. In the early days of the solar system, space was full of these lumps. They kept bumping into one another and sticking together to form larger lumps. After many millions of years, two large round lumps had formed—Earth and the Moon. Another idea is that the Moon formed somewhere else in the solar system. Then, long ago, it wandered close to Earth and was captured. But this seems most unlikely.

Earth's Daughter?

Earth and the Moon could be mother and daughter, some astronomers believe. They suggest that billions of years ago, a huge body the size of Mars crashed into Earth. It knocked off masses of rocky lumps, which were flung into space. In time the lumps came together to form a separate body, the Moon.

Some astronomers have suggested that the Moon might have formed out of material that came originally from Earth. They believe that the material split apart from Earth after Earth was hit by a huge body.

COOLING DOWN

When the great lump of the Moon first formed, it was very hot. The rocks on the surface were so hot that they were liquid. The Moon took millions of years to cool down. As it cooled down, a solid skin, or crust, formed on the surface. Then the crust wrinkled in places, forming a rugged, mountainous landscape. We can still see parts of this old, wrinkled crust in the regions of the Moon that we call highlands.

BOMBARDMENT

Rocky lumps still swarmed in space around the cooling Moon. For millions of years they rained down on the Moon, digging out craters large and small. The biggest lumps created vast circular bowls, or basins, many hundreds of miles across.

LAVA FLOWS

On the surface, the Moon continued to cool down. But on the inside, it began heating up again. Underground rock became liquid and forced its way out through holes and cracks in the crust. Volcanoes erupted all over the Moon. They flooded the low-lying basins with liquid rock, or lava. When the eruptions died down, the lava cooled and formed great flat plains. We call these plains the Moon's seas, or maria.

What the Moon Is Like

The Moon is a rocky body like Earth, but much smaller. And unlike Earth, it has no atmosphere or weather.

The Moon is made up of rock, but it is not a solid ball of rock. It has several different rock layers, just like Earth. The upper layers form the Moon's crust. Beneath the crust is a much thicker layer of a different kind of rock. This is called the mantle. In the center is the core. The rock in the core is hot and is probably partly molten, or liquid.

Scientists learned about the structure of the Moon by studying moonquakes, or shakings in the underground rocks. We call similar events on Earth earthquakes. During a moonquake, shock waves change direction when they hit different rock layers. The *Apollo* astronauts took instruments called seismometers to the Moon to detect moonquake waves. Scientists used readings from the seismometers to learn about the Moon's rock layers.

Above: Diana was the Roman goddess of the Moon, as well as the goddess of hunting. She was also known as Luna, from which we get the word *lunar.* Behind her is the symbol for the Moon, a crescent.

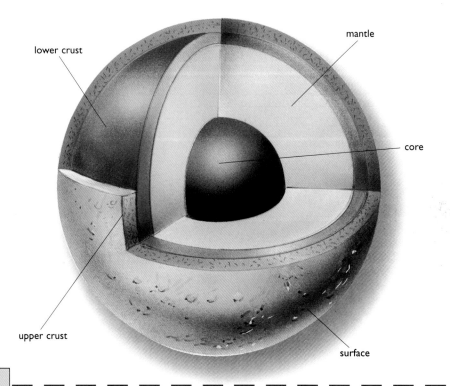

lower crust

mantle

core

upper crust

surface

Left: The Moon is made up of different layers of rock. At the surface is a soft layer a few yards thick, which is much like soil on Earth. Rocks in the upper crust are cracked, and rocks in the lower crust are more solid.

Opposite: *Apollo 11* astronaut Edwin Aldrin sets up a seismometer on the Moon. This instrument detects the waves produced by moonquakes that take place in the underground rocks. About 3,000 moonquakes occur every year.

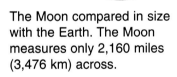

SMALL BUT LARGE

The Moon is only about one-fourth as big across as Earth. It may be small compared with Earth, but it is large for a moon. Of the 60 or more moons in the solar system, only 4 are bigger than the Moon. Because of its size, astronomers sometimes think of the Moon as a small planet. They say that Earth and the Moon together form a double planet.

Because the Moon is relatively small, it has a small mass. And because it has a small mass, it has a low gravity. Gravity is the pull that every heavenly body has on things on or near it. On the Moon, the pull of gravity is only about one-sixth the pull of gravity on Earth. So if you can jump 2 feet (60 cm) high on Earth, you would be able to jump 12 feet (3.5 m) high on the Moon.

The Moon compared in size with the Earth. The Moon measures only 2,160 miles (3,476 km) across.

THE AIRLESS MOON

On Earth, the pull of gravity is strong enough to hold on to the gases surrounding it. These gases form a layer around Earth that we call the atmosphere. But on the Moon, the pull of gravity is too weak to hold on to any gases. The particles in gases travel fast enough to overcome the Moon's pull and escape into space.

With no atmosphere, the Moon is a very different world from Earth. The sky is always dark, even when the Sun is shining. On Earth, the air makes the sky look blue during the day. The Moon is also a silent world, because sounds can only be heard if they have air to travel in. So if you were walking on the Moon with friends and tried talking to them, they wouldn't hear you.

Seen from the Moon, Earth rises above the Moon's horizon. On Earth, we can see white clouds floating in the blue sky. But on our Moon, the sky is black.

Below: Boats lie ashore in a harbor at low tide. In a few hours, they will float high in the water at the next high tide. The Moon causes the tides by pulling at the water in the oceans.

WEATHER REPORT

There is no weather on the Moon. No clouds form, no rain or snow falls, and no dew or frost settles on the ground. Weather occurs on Earth because Earth has an atmosphere that contains water. But the Moon has no atmosphere.

On Earth, the atmosphere helps keep our climate at a comfortable temperature. During the day, it helps spread around the heat from the Sun. During the night, it acts like a blanket and stops too much heat from escaping from Earth back into space.

On the Moon, this process does not happen, and the

MOON DATA

Diameter: 2,160 miles (3,476 km)
Average distance from Earth: 239,000 miles (384,000 km)
Mass: 1/81 Earth's mass
Gravity: 1/6 Earth's gravity
Spins on axis in: 27 1/3 days
Orbits Earth in: 27 1/3 days
Goes through phases in: 29 1/2 days

Moon gets very hot during the day and very cold at night. In addition, days and nights on the Moon—lunar days and lunar nights—are about 14 Earth-days long.

This means that part of the Moon is constantly in sunlight and being heated for two weeks at a time. The temperature of the surface rises to about 260° F (127° C). Then, during the two-week night, the Moon's heat escapes into space. The temperature falls to about −280° F (−173° C), very much lower than any temperatures found on Earth.

The space probe *Clementine* took this picture of the Moon in 1994. It shows the Moon's southern hemisphere (half). In the center is the south pole. *Clementine* found signs that ice is present in this region. Four years later, the probe *Lunar Prospector* also found ice there.

MAKING THE TIDES

Even though the Moon lies a long way from Earth, its gravity still affects Earth. The Moon's gravity pulls at the water in the oceans and makes them rise and fall twice every day. We call this to-and-fro movement of the seas the tides.

The Moon and the tides. When the Moon passes over Earth at (A), it pulls the sea toward it, causing a high tide. This lowers the sea on each side, causing low tides. The Moon also pulls the Earth away from the sea on the opposite side, making a high tide there as well. When the Moon moves on to (B), high tide occurs in that part of Earth.

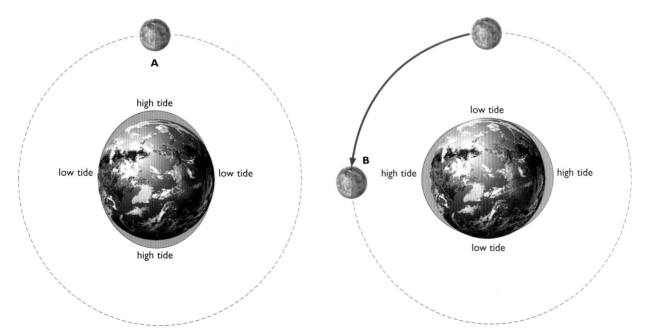

A

high tide

low tide   low tide

high tide

B

low tide

high tide   high tide

low tide

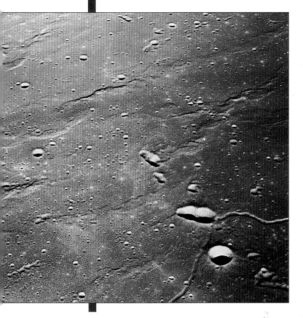

The Moon's Surface

Many different kinds of land features are found on the surface of the Moon. There are vast dusty plains, towering mountains, and craters beyond number.

When we look at the Moon from Earth, we can see two main types of regions on the surface—dark ones and light ones. Early astronomers thought that the surface of the Moon might be like the surface of Earth, made up of seas and land masses. They called the dark regions maria (Latin for seas). They called the light regions terrae (Latin for lands or continents).

We still call the dark regions maria or seas, even though we know they are not made of water. As far as we know, there have never been any watery seas on the Moon. The maria are in fact vast, flat plains. In general, they are lower than the lighter-colored regions, or highlands.

Above: Part of the Moon's Sea of Tranquility. The landscape is very flat and there are only a few small craters. Little channels called rilles snake here and there.

Sea of Moscow

The Far Side

The far side of the Moon looks very different from the side we see from Earth. The main difference is that the far side has no large maria. There is one small mare in the north, the Sea of Moscow. It was named by Russian space scientists, who launched *Luna 3*, the space probe that took the first pictures of the Moon's far side in 1959.

Below: Astronaut Jack Schmitt examines a huge boulder in the highlands around the Sea of Serenity. In the background are the Taurus Mountains.

Early astronomers gave the maria fanciful names, such as the Sea of Showers, the Sea of Serenity, the Sea of Nectar, and the Ocean of Storms. Other names were more sinister, such as the Sea of Crises, the Lake of Death, and the Marsh of Decay.

Many of the maria are circular in shape. The Sea of Crises and the Sea of Serenity are examples. Other maria, such as the Ocean of Storms and the Sea of Nectar, have no particular shape.

THE LUNAR HIGHLANDS

The Moon's highlands are very different from the maria regions. The maria are flat and smooth, but the highlands are very rugged. Mountain ranges stretch for hundreds of miles through many parts of the highlands.

One of the longest ranges is the Apennine Range. It stretches for 400 miles (650 km) around the Sea of Showers. Many of its peaks rise to nearly 20,000 feet (6,000 m).

Another difference between the maria and the highlands is their age. The highlands are much older than the maria. We know this because the highlands are completely covered with craters large and small. Lumps of rock called meteorites have been crashing down on the highlands and producing craters ever since the Moon was formed. But there are only a few small craters on the maria. So they cannot have been bombarded with meteorites for as long.

CRATERS GALORE

Meteorites have rained down on the Moon for millions of years. When they hit the surface, they gouged out pits, or craters. The bigger the meteorite lump was, the larger the crater. When the Moon was young, it was struck by huge lumps, which dug out huge craters. Many craters are more than 100 miles (160 km) across.

There are many kinds of craters on the Moon. Small craters are usually bowl shaped, but large ones are different. A typical large crater has walls that rise high above the surrounding ground. They rose when rocky material was pushed upward after the meteorite hit the surface. On the inside, a large crater drops far below the ground level in tiers, much like the tiers in a sports stadium. In the middle of the crater floor is usually a small mountain range.

Some large lunar craters have low walls, and their floor is very smooth and flat. We call them walled plains. On seas, only the tops of some old craters poke out, because they filled up with lava long ago. We call them ghost craters.

Meteorites rain down on the Moon all the time. The large ones dig out deep craters in the surface. In the past, meteorites weighing millions of tons have dug craters more than 100 miles (160 km) across.

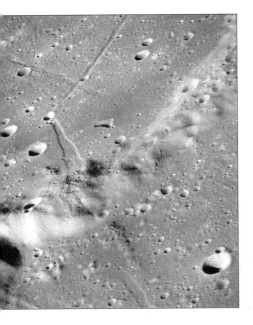

Above: Many small craters dent the surface of the Ocean of Storms. The largest are only a few hundred yards across. Some craters have been made by meteorites, others by tiny volcanoes.

Right: A large crater on the far side of the Moon.

Newton is the deepest crater we know of. Its floor lies more than 5 miles (8 km) beneath the top of its walls.

VOLCANOES AT WORK

Not all craters on the Moon were made by meteorites. Many were made by volcanoes. After a volcano stops erupting, its center collapses. A crater is left behind. This happens on Earth too. Some of the smallest volcanic craters look like dimples and are called dimple craters. Others are found strung out in long lines. We call them crater chains.

Volcanoes have created other features on the Moon. Raised domes appear in some places on the seas. They were probably made by molten rock pushing up toward the surface from below. In other places there are long channels called rilles. They probably carried rivers of lava long ago.

Moon Maps

The Moon is close enough that we can spot many of its features with binoculars or small telescopes. We can spy seas, bays, lakes, craters, and mountains.

The Moon is the only heavenly body we can see clearly. With only our eyes we can see some features, such as the dark-colored maria. We can see many more features through binoculars or a small telescope.

The maps on the next two pages will help you find your way around the Moon's near side and identify the most prominent features on the surface—the maria, the largest craters, and the biggest mountain ranges.

The maps are drawn with north at the top and south at the bottom. They show the Moon as you would see it with the naked eye and through binoculars. But if you look at the Moon through a telescope, you will see it upside down. South will be at the top and north at the bottom. Telescopes always show an upside-down image.

THE EASTERN HALF

The map on the opposite page shows the eastern hemisphere (half) of the Moon's near side. This is the part of the Moon that you see in the night sky at the first quarter phase.

Four main maria show up. Near the eastern edge is the circular Sea of Crises, which measures about 300 miles (480 km) across. Farther west, three larger maria join together—Serenity, Tranquility, and Fertility. The first astronauts to land on the Moon touched down on the southern edge of the Sea of Tranquility in July 1969.

North of the Sea of Serenity, a pair of craters is easy to spot—Aristoteles and Eudoxus. Two mountain ranges ring the sea—the high Caucasus and lower Haemus Mountains. In places the Caucasus range rises up to 20,000 feet (6,000 m).

Following the Terminator

Astronomers call the line between the light and dark parts of the Moon the terminator. During the month, the terminator moves across the Moon's surface from east to west. In telescopes and binoculars, craters and mountains show up well when they are near the terminator because they cast long shadows.

SEA OF COLD

Aristoteles

Eudoxus

Caucasus Mountains

SEA
OF
SERENITY

Apollo 15

Haemus Mountains

SEA
OF
VAPORS

Apollo 17

SEA
OF
CRISES

SEA
OF
TRANQUILITY

Apollo 11

SEA
OF
FERTILITY

Langrenus

Ptolemaeus

Apollo 16

Theophilus

SEA
OF
NECTAR

Alphonsus

Arzachel

Altai Mountains

Petavius

Fracastorius

Walter

Tycho

Clavius

The Eastern Half

Light-colored highlands cover much of this part of the Moon. They are all heavily cratered. Of the maria, the Sea of Crises is the easiest to spot. You can see it a few days after the new moon.

STAR POINT

A new mineral found on the Moon was named armalcolite, after the names of the *Apollo 11* astronauts Armstrong, Aldrin, and Collins.

The Western Half

The western half of the Moon's near side looks quite different from the eastern half because it is covered mainly by dark-colored seas. They all merge together to form the Moon's biggest feature. In the far south is Clavius— one of the Moon's largest craters. It measures 145 miles (230 km) across.

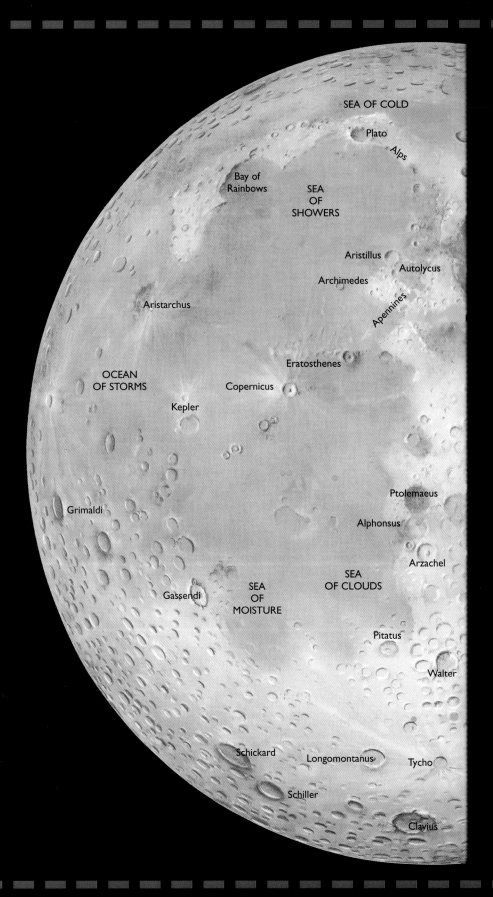

SEA OF COLD

Plato

Alps

Bay of Rainbows

SEA OF SHOWERS

Aristillus

Autolycus

Archimedes

Apennines

Aristarchus

Eratosthenes

OCEAN OF STORMS

Copernicus

Kepler

Ptolemaeus

Alphonsus

Grimaldi

Arzachel

SEA OF CLOUDS

Gassendi

SEA OF MOISTURE

Pitatus

Walter

Schickard

Longomontanus

Tycho

Schiller

Clavius

THE WESTERN HALF

The map on the opposite page shows the western hemisphere of the Moon's near side. This is the part of the Moon that you can see in the night sky at the last quarter phase.

Seas cover most of this hemisphere. In the north is the Sea of Showers. This is the largest of the circular maria. It measures more than 700 miles (1,100 km) across. High mountains border this mare in the north (the Alps) and east (the Apennines). Some peaks in the Apennines soar to over 20,000 feet (6,000 m).

The Sea of Showers overlaps with the largest mare on the Moon, the Ocean of Storms. This sprawling mare has no particular shape but covers an area more than half the size of the United States. In the south, the Ocean of Storms runs into two smaller maria, the Sea of Moisture and the Sea of Clouds.

THE CRATERS

A trio of small craters shows up well on the floor of the Sea of Showers—Archimedes, Aristillus, and Autolycus. There is another arc of three craters farther south—Eratosthenes, Copernicus, and Kepler. Copernicus and Kepler shine brightly during a full moon because of their shining rays or bright streaks.

A whole line, or chain, of craters stands out in the east going south from the center. This chain looks magnificent when it lies on the terminator, both at the first quarter and last quarter phases.

Brilliant Rays

The craters Kepler and Copernicus are not very big. But they are easy to spot, especially at the time of the full moon, because they are surrounded by bright streaks, or rays. Farther south, brilliant rays surround the crater Tycho. They run for hundreds of miles in every direction. Astronomers think that the rays are made up of glassy material thrown out when the craters formed.

Moon Rocks

Several kinds of rock are found on the Moon. They are similar to rocks we find on Earth. In most places, the rocks are covered by slippery soil.

Until the 1960s, no one knew what the Moon was really like. Some astronomers thought that the surface was covered with a deep layer of dust. Anything landing on it would sink and disappear, they said. Then Russian and American probes successfully landed on the Moon, showing that it had a hard, rocky surface.

This paved the way for the *Apollo* astronauts, who began exploring the Moon on foot in 1969. One of their main jobs was to collect samples of Moon rocks and soil and bring them back to Earth so that scientists could find out exactly what the Moon is made of.

On Earth, there are three main kinds of rocks—igneous, sedimentary, and metamorphic. Igneous rocks form when hot, molten rock cools down. Sedimentary rocks form when layers of material accumulate over time. Metamorphic rocks form when another kind of rock changes. On the Moon, there are only igneous rocks. Mostly, Moon rocks are made up of

Below: Two of the rock samples brought back to Earth by Apollo astronauts. Like all other rocks on the Moon, they are volcanic. The bottom one is made up of bits of earlier rocks cemented together, a rock type called breccia.

The *Apollo* Rocks

Between July 1969 and December 1972, *Apollo* astronauts made six landings on the Moon. They explored both the flat lunar plains and the mountainous lunar highlands. They picked up samples of rocks and soil on the surface and also drilled into the ground to get at rocks deeper down. In all, they brought back 850 pounds (385 kg) of Moon rock and soil.

Edwin Aldrin bores into the Moon's surface on the first *Apollo* mission.

minerals very similar to the minerals found in rocks on Earth.

ROCKS OF THE PLAINS AND HIGHLANDS

Different kinds of igneous rocks are found in different parts of the Moon. The rocks of the maria are dark and heavy. They are much like the rock we call basalt on Earth. They formed when molten lava flowed over the maria millions of years ago.

The rocks of the Moon's highland regions are much lighter in color and weight. They are also much older—up to 4.5 billion years old. They were some of the first rocks that formed after the Moon was formed.

SHATTERED ROCKS

When meteorites hit the surface of the Moon, they smash the rocks to pieces. Some hit the surface so hard that they make the rock melt. The melted rock then glues bits of shattered rock together to form big lumps. We call these lumps breccia, which is the name of a similar rock we find on Earth. Lumps of breccia are found all over the Moon, both on the maria and in the highlands.

As meteorites have rained down on the Moon's surface over billions of years, they have smashed many of the surface rocks into tiny pieces. As a result, the surface is now covered with a thick layer of fine material, much like newly plowed soil on Earth. We call this Moon soil regolith. It contains millions of tiny particles of glass. These particles make the soil quite slippery to walk on.

Twelve astronauts left their footprints in the dusty soil that covers most of the Moon's surface. The footprints will last for many years because there is no rain or wind on the Moon to wash or blow them away.

Right: Colored mineral crystals show up in a sample of Moon rock, brought back by the *Apollo 12* astronauts. The rock is a basalt.

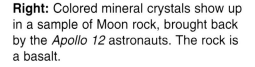

25

Moonwalking

In 1969, human beings planted their footprints in the dusty soil on the Moon. This was perhaps the most exciting adventure in the history of humankind.

President John F. Kennedy first talked about "landing a man on the Moon" in May 1961. This was before any American had even managed to travel in space around Earth. But less than 10 years later, American astronauts were walking on the Moon.

American astronauts began traveling in space in 1962. John Glenn was the first, on February 20. Three other astronauts followed, flying alone in a series of small crafts, or capsules, called *Mercury*. In 1965, U.S. astronauts started flying into space two at a time in the *Gemini* series of spacecraft.

On 10 very successful missions, the *Gemini* astronauts learned new skills. They learned how to guide and control their craft in space. They learned how to rendezvous, or meet up with other craft in orbit. And they learned how to dock with them, connecting two spacecraft together. Also, they began to make spacewalks outside their craft. The astronauts would need all these skills when the time came to try for a Moon landing. The Moon-landing project was called *Apollo*.

Above: An *Apollo* CSM being readied in the Vehicle Assembly Building at the Kennedy Space Center. Overall, it measures about 36 feet (11 m) long. The cone-shaped part at the top is the command module, which houses the crew.

Left: An *Apollo*/Saturn V rocket starts its journey from the Vehicle Assembly Building to the launch pad over 3 miles (5 km) away. With its launch tower, it is carried on top of a huge vehicle with eight caterpillar (crawler) tracks.

APOLLO BUILD-UP

The *Apollo* spacecraft were designed to carry three astronauts. They rode in a capsule called the command module, which was shaped like a cone. Attached to it was a service module, which contained fuel, a rocket motor, and other equipment. The two modules were joined together for most of the time and were known as the CSM (command/service module). The final part of the spacecraft was a spidery-looking craft called the lunar module. This was the part that was designed to land on the Moon.

A huge new rocket was needed to launch the *Apollo* spacecraft to the Moon. It was the Saturn V. It stood 365 feet (111 m) tall and weighed about 3,200 tons (2,900 t). A huge new building was built to assemble the rocket. It was called the VAB, or Vehicle Assembly Building. It became the heart of a new launch site near Cape Canaveral in Florida—the Kennedy Space Center.

In December 1968, three astronauts made their way to the Moon in the *Apollo 8* spacecraft. It was the first time any astronauts had left Earth orbit. They circled the Moon but did not land, and then returned safely to Earth.

"THE EAGLE HAS LANDED"

On July 16, 1969, a Saturn V rocket blasted off the launch pad at the Kennedy Space Center, again heading for the Moon. It carried the *Apollo 11* astronauts Neil Armstrong, Edwin "Buzz" Aldrin, and Michael Collins. Four days later, Armstrong and Aldrin dropped down to the Moon in their lunar module. Its radio identifying name was *Eagle*.

The lunar module touched down safely on the surface of the Moon on the Sea of Tranquility. Armstrong radioed to mission control at Houston, Texas: "Houston, Tranquility Base here. The Eagle has landed."

A few hours later, Armstrong stepped down onto the Moon. "That's one small step for man," he said, "one giant leap for mankind."

Above: The Saturn V rocket blasts off from Cape Canaveral, Florida.

Below: The lunar module descends to the Moon. It fires its rocket engine all the way down. This slows it down so that it can drop gently on the surface on its four spidery legs.

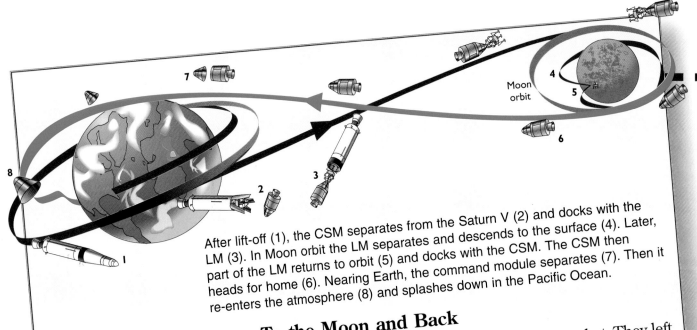

After lift-off (1), the CSM separates from the Saturn V (2) and docks with the LM (3). In Moon orbit the LM separates and descends to the surface (4). Later, part of the LM returns to orbit (5) and docks with the CSM. The CSM then heads for home (6). Nearing Earth, the command module separates (7). Then it re-enters the atmosphere (8) and splashes down in the Pacific Ocean.

To the Moon and Back

The three *Apollo 11* astronauts began their journey on top of a huge Saturn V rocket. They left this behind as they headed for the Moon in the *Apollo* spacecraft, made up of the command/service module (CSM) and the lunar module (LM).

When they reached the Moon, two of the astronauts got into the LM and landed on the surface. They took off in the upper part of the LM and rejoined their colleague in the CSM.

They then headed for home in the CSM. They re-entered the atmosphere in the crew section, or command module, and splashed down in the Pacific Ocean.

EXPLORING THE SEAS

Armstrong and Aldrin explored the Moon's surface for about two and a half hours. They took samples of rocks and soil, and they set up equipment for scientific experiments. Then they took off in the lunar module and rejoined Collins in the *Apollo 11* CSM. Four days later, the astronauts were back safely on Earth.

The next mission, *Apollo 12*, set off in November 1969. It landed on the biggest lunar sea, the Ocean of Storms. One of the astronauts' main jobs was to set up a scientific station. This radioed measurements from instruments back to Earth after the astronauts had left.

Wheels on the Moon

The *Apollo* lunar rover, or Moon buggy, had four-wheel drive. Electric motors drove each of its wheels. Its top speed was about 10 miles (16 km) per hour. The astronauts steered the rover using a kind of joystick.

But the lunar rover was not the first wheeled vehicle to roll over the Moon This honor goes to Russia's *Lunokhod 1*, which landed on the Moon in November 1970. It had eight wheels and two TV camera "eyes."

Above: The lunar rover used by *Apollo 15* astronauts in July 1971

Right: Russia's *Lunokhod 1*, which landed on the Sea of Showers in November 1970.

Apollo 13 set out in April 1970. It was aiming for another site on the Ocean of Storms, but it never got there. An explosion damaged the spacecraft on the outward journey. The astronauts circled the Moon and just barely managed to return to Earth safely. *Apollo 14* headed for the same target, and that mission reached it.

LUNAR ROVING

Three more lunar landings took place. In July 1971, *Apollo 15* set down in the foothills of the Apennine mountain range on the edge of the Sea of Showers. It was a spectacular landing site, with 15,000-foot (4,500-m) peaks towering in the distance. On this mission, the astronauts were able to travel much farther than astronauts on earlier missions. They had a lunar rover, or Moon buggy, for transportation.

The astronauts on the two remaining *Apollo* missions also used lunar rovers. In April 1972, the *Apollo 16* astronauts set down in a highland region southwest of the *Apollo 11* landing site. In December 1972, *Apollo 17* landed in a valley near the Taurus Mountains on the edge of the Sea of Serenity. On board was Harrison ("Jack") Schmitt, the first trained geologist to visit the Moon.

On December 14, Commander Eugene Cernan was the last *Apollo* astronaut to walk on the Moon. "We leave as we came," he said, "and God willing we shall return with peace and hope for all mankind."

The *Apollo 17* command module splashes down in the Pacific Ocean, with its parachutes billowing overhead. The date is December 19, 1972. The Apollo Moon-landing program is at an end.

Moon Base

Someday, astronauts may return to the Moon and set up a base there. One day it could become a spaceport for craft traveling to other planets.

In the first few years of the twenty-first century, the United States and other countries will be hard at work preparing the international space station for its work in space. Before long, they may turn their eyes again to the Moon and make plans for a permanent base there.

A Moon base would be built on the near side of the Moon, which always faces Earth. The base would probably be built using empty rocket fuel tanks. They would be covered with Moon soil for protection. Later, rooms could be hollowed out of the rock, and glass-domed areas built to act as greenhouses for raising crops.

Supplies for a Moon base would have to be ferried from Earth, or from the international space station. But soon Moon base could be self-supporting. This is because there is plenty of water on the Moon, in the form of ice. In 1998, the *Lunar Prospector* probe found ice at both the north and south poles. The ice could supply water for drinking. It could also be converted into oxygen for breathing and fuels for rockets.

Above: This illustration shows what a lunar ferry craft might look like as it sets off for the Moon, carrying cargo for a Moon base. It would have been built in Earth orbit by astronauts living in the space station.

Left: A lunar lander of the future touches down near Moon base. The crew are getting ready to enter the tunnel to the lunar transporter (foreground), which will take them to the base. Other astronauts move in to unload cargo and ready the lander for its next lift-off.

Glossary

atmosphere: the layer of gases around the Earth or another heavenly body

basalt: a dark, heavy kind of rock found on the Moon

breccia: a kind of rock found on the Moon that is made up of older rocks stuck together

crater: a pit in the surface of a planet or a moon

eclipse: what happens when one heavenly body moves in front of another and blocks out its light

gravity: the attraction, or pull, that every heavenly body has on objects on or near it

heavenly body: an object that appears in the sky, such as the Moon or a planet

heavens: space, or the sky

lava: molten (liquid) rock

lunar: having to do with the Moon. A lunar eclipse is an eclipse of the Moon.

lunar rover: a vehicle used by *Apollo* astronauts on the Moon

mare (plural **maria**): a flat plain on the Moon; also called a sea

meteorite: a lump of rock from outer space that hits a planet or a moon

moon: a natural satellite of a planet

Moon buggy: another name for the lunar rover

moonquake: a shaking, or vibration, in the underground rocks of the Moon

orbit: the path in space of one heavenly body around another; for example, of the Moon around Earth

phases: the different shapes of the Moon we see during the month, as more or less of it is lit up by the Sun

probe: a spacecraft that travels from Earth to explore the Moon or other planets

rays: bright streaks around a Moon crater, fanning out from the crater like the spokes of a wheel

regolith: Moon soil

rille: a channel on the Moon, in which lava once flowed

satellite: an object that circles in orbit around a larger body. Earth has one natural satellite— the Moon. It also has many artificial satellites— spacecraft launched by humans.

solar: having to do with the Sun. A solar eclipse is an eclipse of the Sun.

solar system: the Sun and all the bodies that circle around it, including planets and moons

terminator: the line between the light and dark parts of the Moon

tides: the rise and fall in level of Earth's oceans, caused mainly by the gravity, or pull, of the Moon

volcano: a place where molten (liquid) rock escapes to the surface of a planet or moon

Index

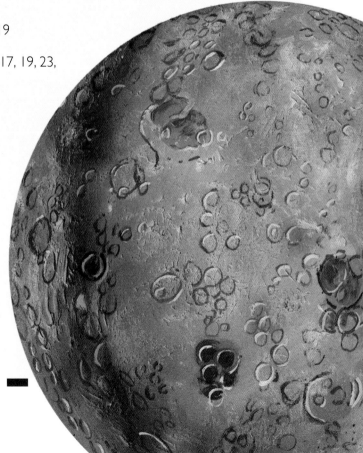

5/01